THE POETRY OF KRYPTON

The Poetry of Krypton

Walter the Educator™

SKB

Silent King Books a WhichHead Imprint

Copyright © 2023 by Walter the Educator™

All rights reserved. No part of this book may be reproduced in any manner whatsoever without written permission except in the case of brief quotations embodied in critical articles and reviews.

First Printing, 2023

Disclaimer
This book is a literary work; poems are not about specific persons, locations, situations, and/or circumstances unless mentioned in a historical context. This book is for entertainment and informational purposes only. The author and publisher offer this information without warranties expressed or implied. No matter the grounds, neither the author nor the publisher will be accountable for any losses, injuries, or other damages caused by the reader's use of this book. The use of this book acknowledges an understanding and acceptance of this disclaimer.

"Earning a degree in chemistry changed my life!"
— Walter the Educator

dedicated to all the chemistry lovers, like myself, across the world

CONTENTS

Dedication v

Why I Created This Book? 1

One - Oh, Krypton 2

Two - Near And Far 4

Three - Secrets Of The Stars 6

Four - Celestial Light 8

Five - Wonder And Love 10

Six - Luminescent Essence 12

Seven - Superman 14

Eight - Tale Of Heroes 16

Nine - Poets And Seers 18

Ten - Krypton's Might 20

Eleven - Thirty-six 22

Twelve - Noble Element Divine 24

Thirteen - Flame Of Curiosity 26

Fourteen - Cosmic Force 28

Fifteen - Muse Divine 30

Sixteen - Progress Is Sown 32

Seventeen - Souls On Fire 34

Eighteen - Unleashing Creativity 36

Nineteen - Sparking Imagination 38

Twenty - Krypton Shine 40

Twenty-One - Gem Of The Sky 42

Twenty-Two - Boundless Imagination 44

Twenty-Three - Whispers Secrets 46

Twenty-Four - Source Of Wisdom 48

Twenty-Five - Symphony Of
Breakthroughs 50

Twenty-Six - Krypton's Core 52

Twenty-Seven - Pure And True 54

Twenty-Eight - Reaching Sky-high 56

Twenty-Nine - Lasting Mark 58

Thirty - Dear Element 60

Thirty-One - Every Leap 62

Thirty-Two - Fabled Birthplace 64

Thirty-Three - Depths Of Time 66

Thirty-Four - Krypton, The Inspiration 68

Thirty-Five - Again And Again 70

Thirty-Six - Catalyst For Exploration 72

About The Author 74

WHY I CREATED THIS BOOK?

Creating a poetry book about the chemical element Krypton was an intriguing and unique endeavor. Krypton, a noble gas, possesses an air of mystery and rarity, which serves as a rich source of inspiration for poets. By exploring the characteristics and symbolism associated with Krypton, I can delve into themes such as isolation, hidden depths, and the search for identity. Poems explore the juxtaposition of Krypton's inertness and the human desire for connection and meaning. This book serves as a creative exploration of science, language, and imagination, merging the realms of science and art in a captivating and thought-provoking manner.

ONE

OH, KRYPTON

In the depths of the cosmos, shining bright,
A gem of the universe, a celestial light.
Born from the stars, so mysterious and rare,
Krypton, the element, beyond compare.

 With atomic number thirty-six, it stands,
A noble gas, in distant lands.
Silent and dormant, in the Earth's embrace,
Krypton's secrets, time cannot erase.

 Its name, a tribute to the ancient lore,
Of Superman's birthplace, forevermore.
Symbolizing strength, hidden and concealed,
Krypton, the enigma, forever revealed.

 Though colorless and odorless, it holds power,
In lighting, lasers, and neon's vibrant shower.

Its electrons dance, in energy's embrace,
Creating a luminescent, ethereal grace.
　From the depths of the night, it casts a glow,
A supernatural aura, for all to know.
In the realm of science, Krypton does reside,
A source of inspiration, impossible to hide.
　Oh, Krypton, element of mystery and might,
A symbol of wonder, in the darkest of night.
As we delve into the unknown, seeking what's true,
Krypton, we honor the light that shines through.

TWO

NEAR AND FAR

In the realm of science, a wonder unfurled,
A celestial light born from the stars,
Krypton, noble gas, with atoms unfurled,
With atomic number thirty-six, it soars.

From distant cosmos, this element arrived,
A symbol of ethereal grace and might,
Its luminescent glow, forever revived,
Illuminating the darkest of night.

In laboratories, its power ignites,
Creating lasers, neon, and light,
A beacon of brilliance, it takes flight,
Guiding us through the realms of twilight.

Superman's birthplace, where heroes reside,
Krypton's legacy, a tale untold,

A planet destroyed, yet it still abides,
In the hearts of those, its story unfolds.
    Oh, Krypton, you inspire us to see,
The beauty in elements and their lore,
With your noble nature, you set us free,
To explore the secrets of the universe's core.
    So let us marvel at Krypton's grace,
A symbol of wonder, both near and far,
As we delve into science's embrace,
May Krypton's light guide us like a star.

THREE

SECRETS OF THE STARS

In the realm of elements, a tale unfolds,
Of Krypton, a substance with secrets untold.
Born in celestial fires, a light so bright,
A symphony of atoms, pure ethereal might.

Krypton, oh Krypton, a celestial grace,
A radiant star-child, in the cosmic space.
With noble gases swirling, in a dance divine,
You paint the heavens, with your luminescent shine.

Through your atomic structure, you create light's glow,
A beacon of brilliance, for all to behold.
Igniting lasers and neon, in electric dreams,
A kaleidoscope of colors, a mesmerizing stream.

Oh Krypton, lost planet, a legacy profound,
Destroyed, but forever, in our hearts you're found.

Your name lives on, in the tales of Superman's birth,
A symbol of hope, in a vast, endless Earth.
From the depths of your essence, inspiration springs,
A freedom to explore, the universe's hidden things.
In your atomic bonds, the secrets of the stars,
Unveiling the mysteries, that lie afar.
Krypton, oh Krypton, your essence so pure,
In the realm of elements, you'll always endure.
A catalyst for wonder, an eternal flame,
In the cosmic symphony, forever your name.

FOUR

CELESTIAL LIGHT

In the realm of celestial light,
A cosmic secret, shining bright,
Krypton, born from distant stars,
Embraced by darkness, yet it mars.
    A noble element, pure and rare,
Its luminescence fills the air,
Within its atomic shell,
Energy dances, we can tell.
    With a glow that's all its own,
Krypton's essence brightly shown,
In gas discharge tubes it thrives,
A spectacle that never dies.
    An element of noble kind,
In Superman's birthplace, we find,

A symbol of hope and strength,
A beacon of justice, at any length.
  Krypton, the catalyst of dreams,
Inspiring explorations, it seems,
Unveiling the wonders of the past,
Through scientific knowledge amassed.
  From distant galaxies it came,
A radiant star-child with no name,
Krypton, a spark of cosmic fire,
Igniting wonder and desire.
  In the darkest nights, it gleams,
Fulfilling our wildest dreams,
A testament to nature's might,
Krypton, a celestial light.

FIVE

WONDER AND LOVE

In the depths of the periodic table's realm,
A mysterious element, rare and overwhelmed.
Krypton, its name, a whisper in the dark,
Symbolizing strength, igniting a spark.

With atomic number 36, it holds power untold,
In noble gases, its place, a treasure to behold.
From light bulbs to lasers, its applications vast,
Krypton shines brightly, a luminescent cast.

Its spectral lines, a symphony in the night,
Unveiling the mysteries, unveiling the light.
Superman's birthplace, a planet so grand,
Krypton, a symbol of hope in a faraway land.

Ethereal grace surrounds this noble gas,
A source of inspiration, destined to surpass.

Its noble nature, a beacon for the wise,
Krypton, the element that never dies.

With noble hearts, we explore the unknown,
Harnessing Krypton's power, a strength we've shown.
A cosmic secret, hidden in the stars above,
Krypton, a catalyst for wonder and love.

So let us marvel at this element divine,
And let Krypton's essence forever intertwine.
In the tapestry of elements, it stands alone,
Krypton, a testament to the universe's throne.

SIX

LUMINESCENT ESSENCE

In the depths of the cosmos, a jewel of light,
Krypton, the element, shining so bright.
With atomic number thirty-six, it does appear,
A symbol of science, discovery, and cheer.

Luminescent glow, a celestial hue,
Krypton, the radiant, captivating the view.
A beacon of wonder, a guide in the dark,
Igniting the curiosity, leaving its mark.

In laboratories, its secrets unfold,
Unraveling mysteries, stories untold.
Noble gas, noble heart, it does possess,
An element of greatness, it does confess.

From Superman's birthplace, its fame may derive,
Yet Krypton's allure goes beyond the superhero's dive.

It symbolizes hope, a world that's long gone,
Yet its legacy persists, forever shining on.
   Through countless experiments, its magic unveiled,
Illuminating discoveries, never to be curtailed.
From lasers to lighting, its uses abound,
Krypton, the catalyst, forever renowned.
   Oh, Krypton, you inspire, you captivate the mind,
With your luminescent essence, endlessly kind.
A symbol of progress, and of dreams to come,
Krypton, the element, forever our guiding sun.

SEVEN

SUPERMAN

In the realm of elements, a sovereign reigns,
A noble gas that in silence remains,
Krypton, the name that commands our gaze,
With powers untamed, it casts its blaze.

A luminescent light, a celestial gleam,
Krypton, the source of a radiant dream,
In lasers it dances, vibrant and bright,
A symphony of photons, a captivating sight.

Neon's vibrant cousin, it shares its might,
Filling the signs that illuminate the night,
In noble gases' realm, it takes its place,
A beacon of brilliance, a celestial grace.

But beyond its glow, a story unfolds,
Of a planet distant, where it once called home,

Krypton, the birthplace of a hero's birth,
A legend's beginning, a tale of great worth.
    Its legacy lives on, in the heart of a man,
Superman, the emblem of truth and plan,
Krypton, the symbol of strength and hope,
A beacon of justice, a hero's oath.
    Inspiring minds, igniting the flame,
Krypton, the muse for science's claim,
A connection to wonders, far and wide,
In realms unexplored, it continues to guide.
    So let us embrace, this element's might,
Krypton, the essence that shines so bright,
A symbol of power, a force to behold,
In its noble presence, stories unfold.

EIGHT

TALE OF HEROES

In the realm of celestial might,
Where stars ignite their cosmic light,
There lies a jewel, a noble gem,
Krypton, the element we must transcend.
   With atomic power deep within,
A legacy of strength it does begin,
Its valiant heart, an unyielding force,
A symbol of power, a prime resource.
   From distant skies, it once did soar,
A beacon of hope, forevermore,
Its presence felt, a radiant glow,
A testament to the wonders we sow.
   Krypton, a catalyst for dreams untold,
A tale of heroes, brave and bold,

In Superman's veins, its essence thrives,
A symbol of hope, where truth survives.
 Through time and space, it lingers on,
A legacy that can never be undone,
Inspiring minds, igniting the flame,
Krypton's power, forever the same.
 So let us marvel at its noble might,
And embrace the wonder, shining bright,
For in Krypton's essence, we find,
A universe of dreams, forever entwined.

NINE

POETS AND SEERS

In a realm distant yet near,
Where stars whisper songs of cheer,
Lies a noble element, bold and bright,
Krypton, bathed in cosmic light.
    From the depths of the universe it came,
A catalyst for wonder, igniting a flame,
A symbol of power, an ethereal might,
Krypton, the beacon in darkest night.
    Its electrons dance in valiant grace,
An atomic structure none can erase,
Noble gas, noble heart, a celestial blend,
Krypton, where dreams and hopes transcend.
    Superman's birthplace, a land renowned,
Where heroes and legends were once crowned,

Krypton, a paradise, majestic and rare,
Its legacy, a tale beyond compare.
   Through the galaxies, its essence soars,
A source of inspiration, forevermore,
It whispers secrets of strength and might,
Krypton, the star that guides us right.
   In the hearts of dreamers, it finds a place,
A symbol of progress, with infinite grace,
Krypton, the muse of poets and seers,
Its allure, eternal, erasing all fears.
   So let our dreams take flight, high above,
Guided by Krypton's everlasting love,
For in its cosmic embrace, we find,
Hope, courage, and a destiny intertwined.

TEN

KRYPTON'S MIGHT

In the depths of cosmic wonders, a treasure lies unseen,
A noble element of mystery, resplendent and serene.
Krypton, the radiant gem, born of celestial embrace,
A symbol of hope and progress, casting its luminescent grace.

From distant realms, it emerged, a catalyst for dreams,
Igniting a legacy of wonder, where hope forever gleams.
In the forge of stars, it was born, a beacon of cosmic might,
A testament to the universe's endless, transcendent light.

Like Superman's birthplace, it shines with bound-

less might,
A world of endless possibilities, eternally burning bright.
Its atomic dance, a symphony of energy and might,
Unveiling the secrets of the cosmos, unlocking the infinite.

Krypton, the eternal flame, inspiring exploration's quest,
Guiding the hearts of dreamers to reach beyond their best.
Its legacy lives on, in the hearts of those who dare,
To dream of a brighter future, where possibilities are rare.

Oh, Krypton, celestial gem, we honor your noble name,
For you are the essence of progress, the spark of endless flame.
In the tapestry of the universe, you forever shall endure,
A symbol of hope and inspiration, forever pure.

So let us raise our voices, in praise of Krypton's might,
And let its luminescent glow guide us through the night.
For in its radiant embrace, we find the strength to roam,
And carry forth its legacy, as we forge a brighter home.

ELEVEN

THIRTY-SIX

In the depths of the cosmos, a legend was born,
A noble element, majestic and forlorn.
Krypton, the name that echoes through time,
A symbol of strength, a legacy sublime.

From the heart of the stars, it came to be,
A beacon of hope, for all to see.
Its atomic number, thirty-six it holds,
A mystery untold, its tale unfolds.

Krypton, the element that birthed a hero,
Superman, the legend, the one we all know.
From the planet's demise, he emerged strong,
A symbol of justice, against all wrong.

But Krypton is more than just a comic book lore,
It's a catalyst for progress, forevermore.
Its power immense, its influence vast,
Igniting dreams, pushing boundaries surpassed.

Through its noble gases, it lights up the sky,
Inspiring explorers, asking them to try.
To reach for the stars, to dream and believe,
For Krypton's essence, it will never leave.

In the realm of science, it holds its place,
A source of endless possibilities, we embrace.
Krypton, the element, eternal and true,
A testament to the wonders the universe can imbue.

So let us remember, as we gaze at the night,
The legacy of Krypton, shining so bright.
A symbol of hope, of dreams yet to come,
Krypton, the element, forever we'll hum.

TWELVE

NOBLE ELEMENT DIVINE

In the realm of science's grand design,
A noble element we find,
Krypton, a name that echoes loud,
A symbol of power, strong and proud.
   Its atomic structure, a mystery untold,
In the depths of the periodic fold,
With electrons dancing, a cosmic ballet,
Krypton's essence, forever at play.
   A gas noble, shining bright,
In the darkness of the night,
A beacon of light, a celestial hue,
Krypton, a symbol of dreams anew.
   From distant stars, its atoms came,
Igniting a fire, a burning flame,

Inspiring heroes, brave and true,
Krypton's legacy, forever to pursue.

In laboratories, its secrets unfold,
Unleashing power, untold and bold,
Scientists marvel, their minds ignite,
Krypton's allure, a brilliant sight.

Poets, too, find solace in its name,
In verses woven, they proclaim,
Krypton's essence, an endless muse,
Igniting passions, lighting the fuse.

Oh Krypton, noble element divine,
In your legacy, we all intertwine,
A symbol of hope, progress, and might,
Guiding us toward a future so bright.

THIRTEEN

FLAME OF CURIOSITY

In the realm where heroes soar,
A power unseen, forevermore.
Krypton, the element of might,
Glimmers in the darkest night.

A catalyst for dreams untamed,
Where imagination is unchained.
Its atomic core, a cosmic spark,
Ignites the fire within the dark.

In laboratories, minds explore,
Unraveling its secrets, to the core.
Science and wonder intertwined,
Krypton's essence, forever enshrined.

From distant realms, it takes flight,
Guiding humanity to new heights.

In comics and tales, its legacy gleams,
Fueling dreams and infinite schemes.

    Oh, Krypton, a symbol of hope,
With strength that helps us cope.
Infinite power, yet fragile still,
A reminder of our human will.

    Embrace its essence, let it ignite,
The flame of curiosity, burning bright.
For in Krypton's legacy, we find,
A universe of possibilities, unconfined.

    So let us explore, and let us dream,
In the realms where Krypton beams.
For within its power, we all can see,
The heroes we were meant to be.

FOURTEEN

COSMIC FORCE

In the vast expanse of cosmic nights,
A noble name forever ignites,
Krypton, a beacon of power and might,
Guiding humanity with its celestial light.
    From the depths of the periodic table,
It emerges with a cosmic fable,
A noble gas, rare and untamed,
In the realms of science, it's acclaimed.
    Krypton, the element of dreams,
Where boundaries fade and hope redeems,
In laboratories and poets' hearts,
Its legacy, forever imparts.
    A symbol of progress, it inspires,
Pushing the limits, reaching higher,

With noble gases, it takes the lead,
Unleashing wonders and planting the seed.

    Oh Krypton, your flame burns bright,
Illuminating the darkest night,
In the hearts of dreamers, you reside,
A spark of passion, never to hide.

    Through your essence, we explore,
Uncharted realms, forevermore,
With every discovery, every new dawn,
Krypton's legacy continues on.

    So let us embrace this noble name,
And let our dreams be set aflame,
For Krypton's power knows no end,
A cosmic force, our souls transcend.

FIFTEEN

MUSE DIVINE

In the realm of stars, where secrets reside,
A glowing gem, Krypton, does confide.
With noble gases, it takes its place,
A symbol of power, in outer space.
    From distant depths, its essence is born,
A legacy of light, forever torn.
Its atomic number, thirty-six,
A beacon of wonder, that all transfix.
    Krypton, a muse for those who dare dream,
Ignites the spark of scientific gleam.
Through laboratories and brilliant minds,
Its influence in science forever binds.
    Unleashing power, unlocking the door,
Krypton's allure, forevermore.

It pushes boundaries, explores the unknown,
A catalyst for progress, brightly shown.
    In the realm of words, poets take flight,
Krypton's inspiration, their guiding light.
Its melody weaves through verses and rhymes,
Igniting passion, transcending times.
    With every stanza, its spirit roams free,
Fueling imagination, for all to see.
A source of might, a muse divine,
Krypton's poetry, a celestial shrine.
    But beyond the realm of science and art,
Krypton shines within every heart.
A symbol of hope, progress, and might,
Guiding humanity through the darkest night.
    So let us embrace this element rare,
And with Krypton's spirit, let us dare,
To dream, to explore, to reach for the stars,
For within us lies the power to go far.

SIXTEEN

PROGRESS IS SOWN

In the depths of the cosmic sea,
Where stars dance with celestial glee,
Lies a noble element, strong and true,
Its name, dear friend, is Krypton, for you.

Born in the heart of a dying star,
With power and might, it traveled far,
A symbol of strength, a beacon of light,
Krypton, the element, shining so bright.

Its atomic number, thirty-six,
A testament to the power it emits,
With noble gases, it proudly abides,
In the realm of elements, it resides.

Krypton, a force that cannot be tamed,
Its influence felt, its power acclaimed,

A catalyst for dreams, it does inspire,
Fueling passions, lifting us higher.
    With its luminescent, ethereal glow,
Krypton's legacy continues to grow,
Pushing the boundaries, exploring the unknown,
Through its essence, progress is sown.
    So let us marvel at Krypton's might,
A symbol of hope, shining so bright,
In the universe vast, it claims its place,
Forever igniting the human race.

SEVENTEEN

SOULS ON FIRE

In the deepest realms of starry skies,
A hidden gem, a noble prize,
Krypton, a realm of ethereal light,
Ignites the hearts with passion's might.

A shimmering specter of rarest hue,
Its essence pure, its power true,
Krypton, a beacon in the cosmic sea,
Unleashing dreams, setting spirits free.

From its atomic heart, a secret lies,
Unveiling truths that mesmerize,
Krypton, a catalyst for boundless thought,
Where innovation and wonder are sought.

With every atom, it dares to explore,
Pushing limits, seeking more,

Krypton, an explorer of infinite space,
Guiding humanity to a higher place.
   Through the tapestry of time, it weaves,
In poets' hearts, its spirit cleaves,
Krypton, an eternal muse of rhyme,
Transcending borders, defying time.
   A force of nature, it holds the key,
To dreams and progress, wild and free,
Krypton, an alchemist of hopes and dreams,
Igniting the flame, forging endless streams.
   Oh, Krypton, embodiment of power and might,
A celestial dance in the darkest night,
Through your essence, we dare to aspire,
To reach for the stars, to set our souls on fire.

EIGHTEEN

UNLEASHING CREATIVITY

In the realms of poets' dreams, Krypton's glow,
A beacon of wonder, its essence aglow.
A noble gas, mysterious and rare,
It dances in shadows, with secrets to share.

From its atomic heart, a tale unfolds,
Of heroes and villains, of legends untold.
With a symbol adorned, a shield of might,
Krypton's legacy shines, in day and night.

Through ink and verse, its power surges,
Igniting imaginations, like cosmic urges.
The poets take flight, their words alight,
Capturing Krypton's essence, so pure and bright.

Its atomic number, a guiding light,
Unleashing creativity, with all its might.

In the minds of artists, it sparks the flame,
A muse that inspires, with no one to blame.
    Krypton, a catalyst, for progress and more,
A catalyst for dreams, to explore and adore.
With every poem penned, a door swings wide,
Revealing new worlds, where dreams reside.
    So let us celebrate, this element divine,
Krypton, the muse, in every poet's line.
For within its atoms, a power untold,
A symphony of words, forever unfold.

NINETEEN

SPARKING IMAGINATION

In the vast expanse of cosmic skies,
A luminescent jewel, Krypton lies.
A noble gas, mysterious and rare,
A muse to poets, beyond compare.

    Krypton, the element of ethereal grace,
Whispering secrets through time and space.
Its atomic number, a mark of distinction,
Igniting minds with poetic conviction.

    Within its atoms, a hidden power,
Inspiring verse, hour after hour.
Words dance upon its spectral glow,
As poets' hearts begin to overflow.

    Krypton, a beacon of creative light,
Guiding souls through the darkest night.

Its presence, a catalyst for inspiration,
Fueling dreams and sparking imagination.
　Through ink and pen, its story is told,
A tale of wonders, both new and old.
It weaves through stanzas, with gentle might,
Elevating thoughts to celestial heights.
　Oh, Krypton! A muse to the human soul,
Unleashing passion, making spirits whole.
In the realm of art, you hold the key,
Unlocking doors to boundless creativity.
　So let us raise our voices high,
To celebrate Krypton's majestic sigh.
For in its essence, we find our voice,
And through its magic, our hearts rejoice.

TWENTY

KRYPTON SHINE

In the realm of poetry, let Krypton shine,
A noble element, a gem so fine.
From the depths of the periodic table it emerges,
With power and grace, its presence surges.
    Krypton, the muse of poets and bards,
Igniting words, like constellations in the stars.
With its atomic number, thirty-six,
It pushes boundaries, explores the unknown's mix.
    A gas so rare, it dances in the air,
Unseen, yet captivating, beyond compare.
It whispers secrets in the poet's ear,
Inspiring verses, making imaginations steer.
    Krypton, the explorer of infinite space,
A catalyst for dreams, a journey's embrace.

It fuels the fires of progress and innovation,
Guiding us forward with boundless inspiration.
  Oh Krypton, muse of rhyme and thought,
Transcending borders, time's constraints it fought.
With noble gases, it forms a symphony,
Unlocking creativity, setting spirits free.
  So let us raise our pens and sing its praise,
For Krypton's light, through poetry, will blaze.
A beacon of inspiration, it will forever be,
Elevating thoughts, setting our spirits free.

TWENTY-ONE

GEM OF THE SKY

In the depths of the cosmos, a force untamed,
Krypton, the element, a name proclaimed.
Born of stars, a celestial birth,
It wanders the universe, exploring mirth.

A beacon of power, a luminescent sight,
Krypton, the explorer, ventures in endless night.
Through nebulae and galaxies it roams,
Seeking the secrets of distant unknowns.

Krypton, the catalyst, ignites dreams anew,
Awakening minds, inspiring breakthroughs.
Its atomic dance, a symphony of light,
Unleashing the potential, igniting the fight.

From laboratories to distant shores,
Krypton's essence, progress it pours.
The spark of innovation, the fuel of thought,
It propels us forward, the dreams we sought.

In the hearts of dreamers, it finds its place,
Krypton, the muse, embracing every space.
Transcending borders, unlocking the mind,
With each whispered rhyme, creativity it finds.

Oh, Krypton, the element, gem of the sky,
With you as our guide, we'll forever fly.
Elevating thoughts, setting spirits free,
In your radiant presence, we find harmony.

TWENTY-TWO

BOUNDLESS IMAGINATION

In the depths of the cosmos, a jewel resides,
A radiant element, where wonder resides.
Krypton, the catalyst of dreams unbounded,
A shimmering muse, where inspiration is founded.

In its noble embrace, creativity awakes,
An ethereal force, where imagination takes,
A flight beyond limits, to uncharted skies,
Where artistry soars, and innovation lies.

Krypton, a beacon, in the vast expanse,
Igniting minds, in a celestial dance,
A luminescent spark, that kindles the soul,
Unveiling secrets, that the universe holds.

Its luminescence whispers, a symphony of light,
Guiding poets and painters, through the darkest night,

Its noble gas essence, a symphony of hues,
Painting dreams on canvases, where brilliance ensues.
   Oh, Krypton, the weaver of dreams untold,
Unlocking the treasures that lie in the fold,
For within your atomic structure, lies the key,
To boundless imagination, and thoughts running free.
   So, let us raise our voices, to this noble gas,
A muse of the heavens, that shall forever last,
In the annals of art, and the human mind,
Krypton, the catalyst, where inspiration we find.

TWENTY-THREE

WHISPERS SECRETS

In the realm where dreams ignite,
A gleaming gem of phantom light,
Krypton, the catalyst of change,
Invisible force, ethereal range.

From the depths of cosmic haze,
It whispers secrets, sets ablaze
The hearts of seekers, minds aflame,
With visions wild, no two the same.

Its atomic dance, a symphony,
Unleashing sparks of creativity,
A muse for minds that dare to soar,
Unlocking worlds unknown before.

As poets weave their tapestry,
Krypton's essence sets them free,
In ink and verse, their spirits rise,
Embodied in celestial skies.

The painters, too, with vibrant strokes,
Find inspiration as Krypton evokes,
Colors unseen, a palette vast,
Creating beauty that will forever last.
　　And in the labs where scientists dwell,
Krypton's secrets they do unveil,
Unlocking mysteries, piece by piece,
Leading humanity to newfound peace.
　　Oh, Krypton, beacon of light,
Guiding us through the darkest night,
Your presence sparks the flames within,
Igniting breakthroughs, let the journey begin.
　　For in your essence, we find our way,
A catalyst for progress, day by day,
Krypton, the element that we adore,
Forever inspiring, forever more.

TWENTY-FOUR

SOURCE OF WISDOM

In the realm of dreams, where stars align,
A catalyst of inspiration, so divine.
Krypton, the element, with powers untold,
A spark of imagination it does hold.

With noble gas it dances, a cosmic ballet,
Igniting minds, lighting the way.
From distant planets to the deepest seas,
Krypton's essence fills the air with ease.

A muse of rhyme, a thought's embrace,
Transcending borders, unlocking grace.
It whispers secrets, in poetic verse,
A symphony of words, a universe.

Through valleys vast and mountains high,
Krypton's essence takes us to the sky.

Unfettered by the chains of time,
It fuels the soul, a rhythm sublime.

    In laboratories, where scientists dwell,
Krypton's presence, a story to tell.
Unlocking doors to realms unknown,
Propelling progress, a victory sown.

    Oh, Krypton, the beacon in the night,
Guiding poets, painters, and minds so bright.
Unveiling secrets, through its gentle glow,
A source of wisdom, forever to show.

    So let us embrace this noble element,
With open hearts and minds, content.
For in its essence, we find our dreams,
Krypton, the muse, igniting our gleams.

TWENTY-FIVE

SYMPHONY OF BREAKTHROUGHS

Krypton, the catalyst of progress,
With powers unseen, it does impress.
A noble gas, in darkness it resides,
Unleashing light, where darkness hides.

Within its atomic structure, a spark,
Igniting ideas, igniting the dark.
A beacon of innovation, it shines,
Guiding scientists through cosmic signs.

In laboratories, minds embrace,
Krypton's energy, their thoughts to chase.
Unlocking secrets, unraveling the unknown,
A symphony of breakthroughs, it has sown.

From lasers to lighting, it leads the way,
Inventing the future, day by day.

Its luminescence, a mesmerizing sight,
Illuminating the path of scientific might.

    Krypton, the muse of poets and painters,
Fueling imagination, setting spirits free.
Through its ethereal glow, inspiration reigns,
Elevating art to heavenly plains.

    With each stroke of the brush, a masterpiece,
Krypton's essence, the artist's release.
In verse and rhyme, it whispers its song,
A muse for the ages, forever strong.

    Oh, Krypton, your power is vast,
Unveiling the treasures of the universe, so vast.
From the depths of the cosmos to the depths of the mind,
You unlock the boundless potential we find.

    Innovation, imagination, and inspiration too,
Krypton, we owe it all to you.
A catalyst for progress, a guiding light,
Forever shining, in our human plight.

TWENTY-SIX

KRYPTON'S CORE

In Krypton's glow, the mind takes flight,
A beacon of inspiration, pure and bright.
From its depths, ideas emerge,
Igniting the flames of genius, they surge.

With noble gas, the spirit soars,
Unlocking secrets, opening new doors.
Imagination dances on its atomic stage,
Unleashing creativity, setting hearts ablaze.

In the realm of art, Krypton's might,
A catalyst for visions, vibrant and light.
Brush strokes flow with ethereal grace,
As painters capture its luminescent embrace.

Scientific minds, in pursuit of truth,
Find solace in Krypton's electrifying youth.

Bathed in its radiance, discoveries are made,
Unraveling the universe's mysteries, unswayed.

Oh, Krypton, your essence divine,
A muse for poets, an elixir of time.
You guide us through the darkest night,
Illuminating the path to knowledge's height.

In your atomic symphony, we find,
A melody that ignites the human mind.
Breakthroughs arise, like stars in the night,
Rooted in the power of Krypton's light.

So let us embrace this element rare,
And let our imaginations take flight, if we dare.
For within Krypton's core, we hold the key,
To unlock the boundless potential of humanity.

TWENTY-SEVEN

PURE AND TRUE

In the realm of science, a gem I see,
A noble gas, shining free,
Krypton, a name that fuels the mind,
Unleashing wonders, one of a kind.

Within its atoms, secrets reside,
A catalyst for breakthroughs worldwide,
Inspiring minds to reach new heights,
Unleashing potential, igniting lights.

With noble grace, it plays its part,
Guiding scientists, brave at heart,
Unraveling mysteries, step by step,
Krypton's essence, they shall not forget.

In laboratories, where dreams ignite,
Krypton's luminescence, a guiding light,

A beacon for artists, poets, and more,
Unleashing creativity, to the core.
　Through brushstrokes and words, it finds its way,
In melodies sung, in colors that sway,
A muse for creators, a spark of fire,
Krypton's luminescence, they aspire.
　Beyond the canvas, beyond the stage,
Krypton unlocks secrets, page by page,
Propelling progress, with every stride,
A symbol of innovation, worldwide.
　Imagination soars, on Krypton's wings,
Unleashing visions, where knowledge springs,
A force of inspiration, pure and true,
Krypton, the element that lights anew.
　So let us embrace, this noble gas,
Elevating art, as the moments pass,
Unlocking treasures, the universe holds,
Guiding humanity, as innovation unfolds.

TWENTY-EIGHT

REACHING SKY-HIGH

In the depths of cosmic night,
Where stars ignite their brilliant light,
There lies a gem, a noble gas,
A secret world, within our grasp.

Krypton, oh mysterious might,
Inert and rare, a captivating sight,
You captivate the artist's eye,
With hues of blue that mystify.

Scientists, in lab coats white,
Unlock your secrets day and night,
Your atomic structure, they explore,
To understand the universe's core.

Philosophers and thinkers wise,
Seek your wisdom, as time flies,

In your noble presence, they find,
A boundless well of knowledge, kind.
 Oh Krypton, beacon in the night,
You guide humanity's relentless flight,
With your atomic number, so high,
You propel progress, reaching sky-high.
 You spark imagination's fire,
Unleash creativity, never tire,
Inspiring minds to dream and soar,
To push beyond what came before.
 Krypton, oh element divine,
Your power, endless and sublime,
We honor you, oh noble gas,
For you inspire, both now and past.

TWENTY-NINE

LASTING MARK

In the heart of the cosmos, a noble gem gleams,
A radiant element, a poet's cherished dreams.
Krypton, dear Krypton, with your ethereal might,
You ignite the imaginations, casting stars in our sight.

Through the eyes of artists, your essence they embrace,
Brushstrokes of brilliance, capturing your grace.
In hues of neon, they paint your cosmic dance,
A masterpiece of wonder, a celestial romance.

Scientists, in awe, unlock your hidden clues,
Unraveling mysteries, with knowledge they infuse.
From your atomic structure, they find inspiration,
Pushing boundaries of knowledge, a relentless exploration.

Thinkers and visionaries, guided by your light,

Illuminate the path, in the darkest of night.
Invention and progress, born from your spark,
Propelling humanity forward, leaving a lasting mark.

    Krypton, oh Krypton, your power knows no bounds,
A beacon of inspiration, where potential resounds.
From the depths of imagination to the realms of science,
You, dear element, ignite our souls' reliance.

    So let us celebrate, the wonders you bestow,
Krypton, dear Krypton, forever may you glow.
In the hearts of dreamers, your legacy will endure,
A testament to the magic of the universe, pure.

THIRTY

DEAR ELEMENT

In the depths of the universe, where stars reside,
There lies a noble element, with grace and pride.
Krypton, the catalyst of inspiration's fire,
Ignites the souls of dreamers, taking them higher.

Oh, Krypton, thou art a muse for poets' quill,
Whose ink spills forth with words that make hearts thrill.
Thy luminescence, a beacon in the night,
Guides painters' hands, creating strokes so bright.

A noble gas, thy power is unseen,
Yet it fuels the fire of scientific dreams.
In laboratories, minds are set ablaze,
Unraveling mysteries, unlocking cosmic haze.

Krypton, thou art the key to boundless potential,
A catalyst for progress, ever essential.

Thy very presence sparks the human mind,
To reach for the stars, leaving limits behind.

    Oh, Krypton, thou art a gift from the divine,
Unveiling secrets, expanding the confines.
Thy electrons dance, a symphony so grand,
Revealing the wonders of this vast, cosmic land.

    So let us celebrate this noble element,
For it holds the power to fuel our ascent.
Krypton, the muse that sets our spirits free,
We owe our inspiration, dear element, to thee.

THIRTY-ONE

EVERY LEAP

In a realm beyond our sight,
Where stars ignite the darkest night,
Lies a noble element, shining bright,
Krypton, a muse that fills us with delight.

From the depths of the periodic table,
Krypton emerges, a celestial fable,
Its atomic number, a cosmic sign,
Unleashing powers that forever bind.

Krypton, the beacon of inspiration,
A catalyst for human innovation,
It unlocks the doors of imagination,
Igniting creativity, without hesitation.

With its noble gas, it whispers secrets,
Guiding scientists on their quests,

Unlocking mysteries, revealing truths,
Krypton, the muse that never snoozes.
  Through the lens of its ethereal glow,
Scientists explore, they seek to know,
From the depths of the universe vast,
Krypton unveils the secrets amassed.
  Progress propelled by its atomic might,
Humanity soars to new heights,
In laboratories, minds take flight,
Krypton's power ignites the light.
  From comic tales to scientific lore,
Krypton's influence forevermore,
A symbol of potential, a guiding star,
Shaping the human experience, near and far.
  So let us celebrate this element divine,
Krypton, the muse that helps us shine,
With every discovery, every leap,
Krypton's legacy, forever we'll keep.

THIRTY-TWO

FABLED BIRTHPLACE

In the realm of the unknown,
Where secrets hide, yet to be shown,
There lies a spark, a cosmic light,
Krypton, the muse that ignites.

    A noble gas, tranquil and rare,
Its power whispers in the air,
A beacon of potential, untamed,
In scientific minds, it is named.

    Oh Krypton, you inspire and guide,
Unveiling truths we cannot hide,
With spectral lines, a vibrant hue,
You paint the skies, revealing the new.

    From distant stars, you journeyed here,
Unleashing wonders, crystal clear,

In laboratories, minds collide,
With your essence, they're unified.
 From Superman's fabled birthplace,
To scientific feats in space,
You symbolize the human mind,
Unearthing treasures, undefined.
 In art and music, you reside,
A source of passion deep inside,
Your energy, a creative fire,
Fueling imaginations to aspire.
 So let us celebrate your might,
Krypton, a catalyst of light,
In science, art, and dreams untold,
Your essence shines, forever bold.

THIRTY-THREE

DEPTHS OF TIME

In the realm where stars are born,
There lies a noble gas forlorn.
Krypton, a muse for creators,
Ignites imagination's embers.

From the depths of cosmic seas,
It whispers secrets with a breeze.
A symbol of innovation's might,
Krypton fuels the dreamer's flight.

With its luminescent glow,
It sparks ideas that freely flow.
Crafting worlds in minds untamed,
Krypton's essence leaves no dream unclaimed.

Its atomic structure, noble and rare,
Empowers minds to venture where

No others dare to tread,
Exploring realms where wonders spread.
    Oh, Krypton, guide humanity's quest,
Illuminate paths where knowledge rests.
In laboratories, minds ignite,
Unleashing progress, shining bright.
    From the artist's brush to the scientist's pen,
Krypton's wisdom flows again.
Unlocking secrets of the universe vast,
It fuels our dreams, transcending the past.
    Though lost in the depths of time,
Krypton's legacy forever will shine.
A beacon of inspiration, ever bright,
Guiding us towards infinite light.

THIRTY-FOUR

KRYPTON, THE INSPIRATION

In the depths of the cosmos, a jewel gleams,
A rare and noble element, Krypton, it seems.
With atomic structure, so noble and strong,
It inspires imaginations, propelling us along.

Krypton, the beacon of limitless might,
Unleashing our dreams, like stars in the night.
A muse for inventors, poets, and scribes,
Igniting creativity with its ethereal vibes.

From science's lab to the artist's canvas,
Krypton's power transcends boundaries, boundless.
It fuels our inventions, pushes us to explore,
Unlocks mysteries, leaving us wanting more.

In laboratories, its secrets unfold,
Revealing the wonders that were once untold.

Through its spectral lines, we gain insight,
Into the universe's secrets, hidden from sight.
    Krypton, the catalyst for human progress,
A catalyst for minds, forever undress.
It breathes life into ideas, ignites our will,
To reach for the stars and climb every hill.
    In its rarity, Krypton holds a key,
To unlocking the potential, in you and in me.
A legacy it leaves, a light that won't fade,
Krypton, the inspiration, in every shade.

THIRTY-FIVE

AGAIN AND AGAIN

In the realm of elements, behold!
Krypton, a name that echoes bold.
A catalyst of progress, it shines,
A source of inspiration, divine.
   From distant stars, its atoms soar,
To Earth, they journey, evermore.
In noble gases, it finds its place,
With power and grace, it leaves a trace.
   Krypton, the muse of inventors' dreams,
Igniting minds with brilliant beams.
Through its glow, new wonders unfold,
Invention's tale, forever told.
   In laboratories, it takes its stand,
Guiding seekers with a steady hand.

Scientists delve into its core,
Unlocking secrets, forevermore.
    From neon lights to laser beams,
Krypton's luminescence gleams.
A beacon of knowledge, it shines bright,
Illuminating the darkest night.
    Oh Krypton, fuel our imagination,
Ignite the fires of exploration.
With every discovery, we soar high,
Guided by your celestial sigh.
    Invention, science, art, and more,
Krypton's touch, forever we adore.
A symbol of progress, it will remain,
Inspiring minds, again and again.

THIRTY-SIX

CATALYST FOR EXPLORATION

In the realm of science, a gem does gleam,
Krypton, the element, a radiant dream.
With atomic number thirty-six,
Its noble nature, it does transfix.

A beacon of knowledge, it does guide,
Unraveling mysteries, side by side.
Fueling inventions, igniting the spark,
Krypton's power leaves its mark.

A light in the darkness, it does shine,
Illuminating the deepest confines.
Through its glow, seekers find their way,
In the darkest night, Krypton holds sway.

Imagination, it does inspire,
As seekers reach higher and higher.

Unlocking secrets, hidden from view,
Krypton's essence, forever true.

    Oh, Krypton, the luminescent star,
A symbol of progress, both near and far.
A catalyst for exploration and quest,
Fueling the fire, that burns in our chest.

    In laboratories, its wonders unfold,
Revealing stories, yet untold.
Krypton, the element, steadfast and true,
With each discovery, it breaks through.

    So let us celebrate this noble gas,
A source of light, as time does pass.
Krypton, the element, forever bright,
In our hearts, it takes flight.

ABOUT THE AUTHOR

Walter the Educator is one of the pseudonyms for Walter Anderson. Formally educated in Chemistry, Business, and Education, he is an educator, an author, a diverse entrepreneur, and he is the son of a disabled war veteran. "Walter the Educator" shares his time between educating and creating. He holds interests and owns several creative projects that entertain, enlighten, enhance, and educate, hoping to inspire and motivate you.

Follow, find new works, and stay up to date
with Walter the Educator™
at WaltertheEducator.com

www.ingramcontent.com/pod-product-compliance
Lightning Source LLC
LaVergne TN
LVHW051958060526
838201LV00059B/3715